Lincoln Bernardo de Souza
Sergio B. de Oliveira
Danns P. Barbosa

Photocatalytic reduction of nitrate in water

Lincoln Bernardo de Souza
Sergio B. de Oliveira
Danns P. Barbosa

Photocatalytic reduction of nitrate in water

Synthesis and evaluation of mono- and bimetallic Ag, Cu, Pd and Sn catalysts supported on titanium dioxide

ScienciaScripts

Imprint

Any brand names and product names mentioned in this book are subject to trademark, brand or patent protection and are trademarks or registered trademarks of their respective holders. The use of brand names, product names, common names, trade names, product descriptions etc. even without a particular marking in this work is in no way to be construed to mean that such names may be regarded as unrestricted in respect of trademark and brand protection legislation and could thus be used by anyone.

Cover image: www.ingimage.com

This book is a translation from the original published under ISBN 978-613-9-72100-9.

Publisher:
Sciencia Scripts
is a trademark of
Dodo Books Indian Ocean Ltd. and OmniScriptum S.R.L publishing group

120 High Road, East Finchley, London, N2 9ED, United Kingdom
Str. Armeneasca 28/1, office 1, Chisinau MD-2012, Republic of Moldova, Europe
Printed at: see last page
ISBN: 978-620-7-94710-2

ACKNOWLEDGEMENTS

I owe part of this master's work to my dear wife, who has always supported, encouraged and encouraged me in this endeavour, always welcoming me with open arms after a weekend working in the lab and with whom I have shared some great moments during this process.

To my parents and brother, who were also very important in carrying out this research, giving me strength even at a distance.

To Professor Sérgio Botelho for his friendly, ethical and respectful behaviour, worthy of a born educator, who made me want to continue in the field of catalysis even after this stage was over.

To Professor Danns Barbosa and Professor Ítalo Fernandes for their guidance and help in carrying out the experiment.

To the chemist and friend Lucas Gaio for his willingness to ask to do some analyses at UnB for the research.

To Maria Margarida Machado and her family for their assistance and help here in Goiânia.

To the Federal Institute of Brasilia, in the form of Professor Sérgio Gomes, for his understanding, respect and cordiality when I was unable to attend many IFB activities due to my master's research.

To the Fundação de Amparo à Pesquisa do Estado de Goiás for the financial support of the research.

To everyone who was supportive in any way and helped me achieve this goal.

SUMMARY

Nitrate is one of the most commonly found compounds in watercourses due to the infinity and diversity of existing sources, whether natural or man-made. Its excess can lead to various health and environmental problems, but cheap and efficient ways of treating water contaminated with nitrate are not yet a reality. One of the ways of remediating aquifers or water sources contaminated with nitrate or nitrite, which can add low cost and efficiency, is the use of photocatalysis, which uses ultraviolet (UV) or visible radiation as an energy source for the degradation of nitrate and nitrite. Focusing on this form of treatment, new photocatalysts were synthesised for the reduction of nitrogenous species such as nitrate and nitrite to molecular nitrogen. Two different materials were used to support the mono- and bimetallic catalysts: titanium dioxide nanowires (TiO_2) and titanium dioxide doped with nitrogen and carbon ($NCTiO_2$). The metals were deposited on the surface of the supports using the impregnation method. Silver (Ag) and palladium (Pd) were used to produce the monometallic catalysts, while the bimetallic catalysts were obtained by depositing copper (Cu) and tin (Sn) onto the former. The photocatalytic tests were carried out with a solution of potassium nitrate and formic acid (electron donor) prepared in the laboratory, under a flow of nitrogen, at a controlled temperature, using a UV radiation source. Both the Nfios support and the mono- and bimetallic catalysts supported on it did not obtain results comparable to its precursor (Degussa P25, 80% anatase, 20% rutile), and the metallic impregnation practically did not improve the photocatalytic activity of the material. The $NCTiO_2$ support proved to be very effective for the photocatalytic reduction of nitrate, with results similar to those of its precursor. Metal deposition of 5% Pd and 2% Sn in relation to the total mass on $NCTiO_2$ led to complete degradation of the initial nitrate three hours earlier than the best result found with P25 (P25-Ag1%), and selectivity to N_2 of 97%, the highest among the catalysts studied. The influence of the concentration of the reaction's electron donor (formic acid), the radiation source and the importance of activating the catalyst with H_2 before the test were observed. The ability to reconstitute the active nitrate and nitrite reducing sites of the catalyst was also assessed, which remained active and highly selective to N_2 after being reused five times.

KEY WORDS: Nitrate. Nitrite. Photocatalysis. Titanium dioxide. Bimetallic catalysts.

SUMMARY

CHAPTER 1

INTRODUCTION

Nitrogen is one of the most important chemical elements for the birth, development and maintenance of life. It is from the incorporation of this element that the macromolecules, deoxyribonucleic acid (DNA) and ribonucleic acid (RNA), responsible for the existence of life as it is on planet earth, originated.

The survival of living beings depends above all on available sources of nitrogen. In the case of humans and other heterotrophic beings, the incorporation of nitrogen compounds occurs through the consumption of autotrophic organisms that are able to incorporate them into their structure. In this way, plants, which are autotrophic beings, need nitrogen compounds to grow, which they find in the soil as nitrate (NO_3^-), which is the form they can absorb. The other forms of nitrogen that are very important for plant development are nitrite (NO_2^-) and ammonia, which can be oxidised to produce the nitrate they need.

Ammonia is easily volatilised in neutral and alkaline environments, and is present in water and soils generally in the form of the ammonium ion (NH_4^+). Nitrite is not very stable and can be considered a by-product of the process of ammonium nitrification and nitrate denitrification, mainly found in groundwater contaminated with anaerobic and reductive nitrogen compounds. The nitrate molecule is the most stable of the inorganic nitrogen compounds in nature, and nitrite and ammonium contained in soils and waters in the presence of oxygen and\or micro-organisms are readily transformed into nitrate. Therefore, most cases of water and soil pollution by nitrogenous species are due to nitrate, caused by excessive fertilisation and fertilisation of crops, inefficient or non-existent treatment of domestic, rural and industrial effluents, inadequate disposal of solid waste (dumps), or atmospheric pollution from the use of fossil fuels.

As well as being harmful to human health, excess nitrogen compounds have a considerable environmental, social and economic impact, especially in lenticular and lacustrine water environments, and can lead to the phenomenon known as eutrophication, which practically depletes the biodiversity existing in the area and can lead to the appearance of algae and cyanophyceous bacteria that produce substances toxic to humans (RESENDE, 2002).

In its latest report on the quality of water for safe human consumption, the World Health Organisation (WHO) recommended maximum permissible values (VMP) of 11 mg N-NO3$^-$ .L^{-1} for nitrate (as nitrogen) or 50 mg NO3$^-$ .L^{-1} and for nitrite 0.9 mg N-NO2$^-$ .L^{-1} (as nitrogen) or 3 mg NO2$^-$ .L^{-1} (WHO, 2011).

The National Legislation that regulates the VMP for human consumption of water is the Ministry of Health Ordinance No. 2,914 of 2011, which practically follows the WHO recommendations, with slightly more restrictive values, 10 mg N-NO3$^-$ .L^{-1} and 1 mg N-NO2$^-$ .L^{-1} (MINISTÉRIO DA SAÚDE, 2011).

Taking into account these two documents, which provide guidelines for estimating the quality of water for human consumption, studies were carried out in the state of Goiás to estimate contamination by nitrogen compounds on a local scale, which may originate from sectors that are very present in this region, such as agro-industry (slaughterhouses, alcohol plants, etc.), agriculture and animal husbandry. In addition to the fact that there are still large numbers of cesspits and septic tanks in homes and industries in this state, as there are throughout Brazil, to dispose of effluent.

Barbosa and Araújo (2009) analysed samples of raw water upstream of water treatment plants (WTPs) in various cities in the southern mesore- gion of Goiás and found nitrate values above those considered safe for human consumption by the WHO (11 mg N-NO3$^-$ .L^{-1}) in Quirinópolis (307.8 mg N-NO3$^-$ .L^{-1}) and Morrinhos (17 mg N-NO3$^-$ .L^{-1}). They concluded that this pollution is related to anthropogenic sources, since the geological units in this region do not have nitrate or nitric compounds in their chemical composition and because there are crop, pig and cattle confinement facilities on the banks of the watercourses analysed.

Campos and Rohlfs (2010) analysed public supply wells in the city of Águas Lindas de Goiás between 2009 and 2010 and found nitrate contamination, with levels above those recommended by the WHO only found in one well sample, although several samples were found to have nitrate levels close to the value considered safe for consumption.

Carvalho and Siqueira (2011) analysed water samples from the Meia Ponte River in the metropolitan region of Goiânia between 2004 and 2008 and found no values above those recommended by the WHO in relation to nitrate, but samples with higher values were detected for nitrite and ammoniacal nitrogen ($NH4^+$).

These studies show that there is a large input of nitrogen compounds into the state's groundwater and surface water, estimates and values that could be even higher today due to the lack of public policies to prevent this type of pollution and the characteristics of Goiás' soils, which have a high permeability to water, which brings more and more contaminants, especially into underground aquifers.

Barbalho and Campos (2010) reported in their study on the natural vulnerability of soils in the state of Goiás to contamination by vinasse used in fertigation that these soils have very high to moderate vulnerability to contamination, mainly by nitrates, which have high solubility and the capacity to percolate and leach into the soil, making this agricultural technique a possible source of surface and groundwater pollution.

In view of this contamination situation, it is necessary to look for alternatives for the treatment of nitrate and nitrite in drinking water, since many people, not only in rural areas but also in towns and cities, usually small ones, use raw water from underground aquifers and surface watercourses for their daily consumption.

According to Barrabés and Sá (2011), nitrate is normally treated by reverse osmosis, ion exchange and electrodialysis processes. These only remove the nitrate from the water and concentrate it in filters or membranes, leading to the formation of brines and sludge that subsequently need to be treated. There is also biological treatment which degrades $NO3^-$ producing nitrogen ($N2$), but this process can lead to the production of other nitrogen oxides and its management is more complex, since the pH, temperature and nutrients in the environment must be controlled in order to achieve high efficiency.

Heterogeneous catalysis is one of the ways to achieve the degradation of nitrate and nitrite to molecular nitrogen with the production of low amounts of by-products. Several studies have explored catalytic reduction as a way of degrading $NO3^-$ using hydrogen (H_2) as a reductant, but the formation of a considerable amount of ammonium as a by-product is one of the disadvantages of this type of treatment (BARBOSA et al., 2013; BARRABÉS & SÁ, 2011; EPRON et al., 2001; SÁ et al., 2005; SÁ & ANDERSON, 2008).

In recent years, photocatalytic reduction has been developed as a way of degrading nitrate in order to obtain low by-product formation and lower operating costs, since there is no need to inject H_2 into the system (GAO et al., 2004; SÁ et al., 2009; SHAND & ANDERSON, 2013; SOARES et al., 2014; YANG et al., 2013; ZHANG et al., 2005). Barrabés and Sá (2011) consider that the photocatalytic treatment of nitrate is the most promising system for dealing with this problem, as it destroys the pollutant with the production of non-toxic N2, uses only solar radiation as an energy source, has low operational complexity and could possibly be applicable in isolated and remote locations. However, the obstacle to developing this technique is the unwanted formation of ammonia in the process, which is also a water pollutant and can be oxidised by the environment to form nitrate again. The best results described in the literature are obtained using small molecules of carboxylic acids as reductants, which improves the operationalisation of the process compared to catalytic reduction, as the oxidation of these electron donors provides the solution with CO_2 and H_2. The former buffers the medium leaving it at an acidic pH, which discourages the formation of ammonium, and the latter helps to increase the activity of the catalyst used. Furthermore, as these carboxylic acids are common by-products of advanced oxidative treatments of water contaminated with organic compounds, photocatalytic reduction can be used later to degrade remaining nitrates and nitrites (SHAND & ANDERSON, 2013). Formic acid acts as a reducing agent by reacting with the gaps created (h^+) in the valence band of the catalyst support due to the electronic excitation promoted by the ultraviolet (UV) radiation coming from the lamp, illustrated in the diagram in figure 1. The reduction of NO_3^- and NO_2^-, which is formed during the reaction, takes place in the conduction band, where they react with the excess electrons located there.

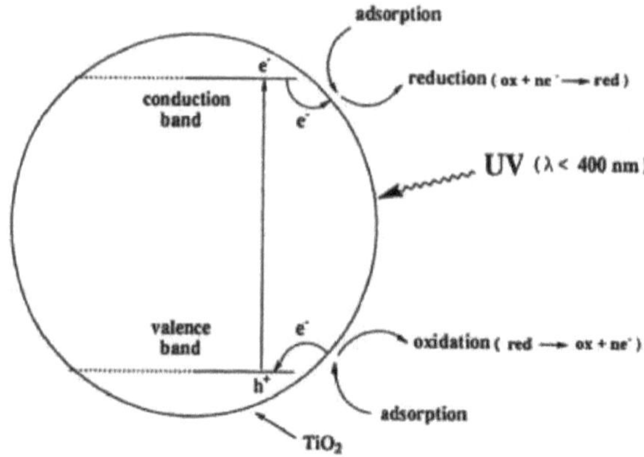

Figure 1 - Schematic of the photocatalytic process on catalysts supported with TIO2 (SHAND & ANDERSON, 2013). With adaptations.

The choice of titanium dioxide (TiO_2) as a support for mono- and bimetallic catalysts is related to its various characteristics such as: low susceptibility to CO_2 poisoning, ability to absorb UV light, high thermal, chemical and biological resistance, low toxicity (LD50 < 10.000 mg/kg body weight of rats (Opinion of the scienti- fic committee on comestic products and non-food products intended for consumers, 2000)), low cost, insolubility in water, photostability to corrosion and ease of generating the electrons and gaps necessary for the redox process to take place.

This work consisted of synthesising new mono- and bimetallic catalysts supported on two different TiO2-based materials (nanowires and doped with nitrogen and carbon) and carrying out photocatalytic tests to check the performance of these catalysts in the degradation of nitrate in aqueous solution. The metals silver (Ag) and palladium (Pd) were used to produce the monometallic catalysts and the metals copper (Cu) and tin (Sn) were deposited on the Ag and Pd-based catalysts for the bimetallic catalysts. After selecting the catalyst with the best results, various tests were carried out to observe the influence of the concentration of formic acid, which is the electron donor in the reaction, the radiation source (UV-C and UV-A), the activation and reuse of the catalyst in relation to the photocatalytic activity and the selectivity of the products (nitrite, ammonium and molecular nitrogen).

CHAPTER 2

DEVELOPMENT

1. METHODOLOGY

The main aim of the research is to develop mono- and bimetallic catalysts supported on titanium dioxide nanowires and titanium dioxide doped with nitrogen and carbon for the photocatalytic reduction of nitrate, achieving high selectivity for N_2. The work therefore has a quantitative aspect, as the aim is to quantify the degradation of these water pollutants, and an experimental and explanatory aspect, as it will be developed through experimental tests and characterisation of the catalysts that will explain the results obtained.

The work is divided into four experimental stages:

- Synthesis of TiO_2 supports, preferably in the anatase crystalline phase (nanowires and titanium dioxide doped with nitrogen and carbon);
- Deposition of metals on the surface of supports for the preparation of mono- and bimetallic catalysts;
- Experimental tests on the photocatalytic reduction of nitrate and nitrite in water with the supports and catalysts.
- Parameter variations in the experimental tests for the catalyst that obtained the best results in terms of photocatalytic activity and product selectivity.

1.1 preparation of the TiO_2 substrates

1.1.1 synthesis of $tio2$ nanowires

The nanowires were synthesised by the solvothermal method using 1.5 g of commercial TiO2 (Degussa P25, 80% anatase and 20% rutile) as a precursor, added to 50 mL of ethanol solution and 10 mol.L NaOH aqueous solution^{-1} in a ratio of 1:1 (v/v), subjected to a temperature of 170 °C in an autoclave for 24 hours. The system was allowed to cool to room temperature, and the solid obtained was washed several times with dilute HCl (0.1 mol.L^{-1}) and deio- nised water, until the pH of the wash solution was close to 7. The washed solid was then dried at 60 °C for 12 hours in an air oven (WEN et al., 2005). This support was called Nfios.

1.1.2 NITROGEN AND CARBON DOPED   TiO2 SYNTHESIS

Titanium dioxide doped with nitrogen and carbon was produced using 4 g of TiO2 (Degussa P25) dispersed in a 10 ml solution of 30 % ammonium hydroxide (NH4OH) and 1-butanol in a 1:1 (v/v) ratio. The suspension was placed in an autoclave at 100 °C for 4 hours. It was allowed to cool to room temperature and the product obtained was dried at 105 °C for 24 h in an air oven (DOLAT et al., 2012). This support was named NCTiO2.

1.2   DEPOSITION OF METALS ON SUPPORTS

Metals were deposited on the surface of the supports using the impregnation method to produce monometallic catalysts. The bimetallic catalysts were made by successive impregnation of the monometallic catalyst.

1.2.1   PREPARATION   OF   MONOMETALLIC   CATALYSTS   BY IMPREGNATION

This method consisted of dispersing the support in an aqueous solution and then adding the solution of the metal salts to be deposited, drop by drop, while stirring for 4 hours. The suspension was evaporated with periodic stirring at 80 °C for 12 hours. The material obtained was dried in an oven at 90 °C and calcined at 300 °C to destroy the complexes that can be formed between the metals and the anions of the precursor salts (chlorides, nitrates and sulphates) near the surface of the support. In this work, monometallic catalysts were produced with the deposition of two different metals, silver (Ag) and palladium (Pd), in a proportion of 5 % of metal in relation to the total mass of the catalyst (mt). For ease of understanding, the monometallic catalysts were named Nfios-X, and NCTiO2-X with X = Pd and Ag, Nfios and NCTiO2 being the corresponding TiO2 supports used (BARBOSA et al., 2013; DODOUCHE et al., 2009).

1.2.2 PREPARATION OF BIMETALLIC CATALYSTS BY SUCCESSIVE IMPREGNATION

Successive impregnation is a repetition of the methodology presented above, but instead of using the support as a precursor, the monometallic catalyst is used. The proportion of the second metal deposited for the synthesis of the bimetallic catalysts was 0.5, 2 and 4 % (mt), using the metals copper (Cu) and tin (Sn) (BARBOSA et al., 2006),

2013; DODOUCHE et al., 2009). The catalysts obtained were identified as Support-X%-Y%, where X refers to the first metal deposited (Ag or Pd), Y to the second metal (Cu or Sn) and % is the mass proportion of metal in relation to the total (mt). Mono- and bimetallic catalysts supported on TiO2 (Degussa P25), the precursor of the supports, were also synthesised in order to compare the results with those of catalysts supported on NCTiO2 and Nfios.

1.3 EXPERIMENTAL TESTS ON THE PHOTOCATALYTIC REDUCTION OF NITRATE AND NITRITE

Before each test was carried out, 128 mg of catalyst was activated in the photocatalytic reactor under a flow of H2 (100 mL.min^{-1}) for 1 h to generate reduced metal active sites, the monometallic Pd catalysts were heated to 150 °C and the others to 200 °C, according to the temperature programmed reduction (TPR) of each material (EPRON et al., 2001; GAO et al., 2004; SÁ & ANDERSON, 2008; SÁ et al., 2009; YANG et al., 2013; ZHANG et al., 2005). The system was cooled to room temperature and 200 mL of a solution of 100 mg NO_3^- .L^{-1} and 10 mmol.L^{-1} of formic acid was added, degassed with N2 (100 mL.min^{-1}) for 10 min.

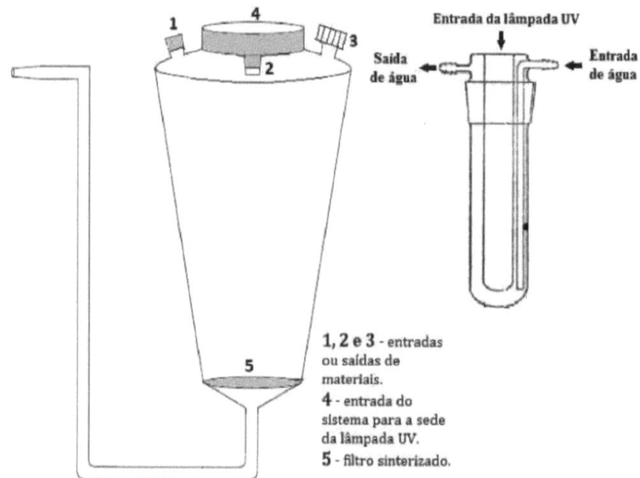

Figura 2. Photocatalytic test reactor (left) and UV lamp seat (right).

The tests were carried out in the photocatalytic reactor illustrated in Figure 2, which was built in such a way as to minimise the effect of temperature and dispense with the use of electrical equipment for stirring the system, since this is done by the flow of N2, which also has the purpose of dragging oxygen from the reaction medium. An Osram HNS 16W G5 lamp was used to emit UV-C radiation (200 - 280 nm), N2 flow (150 mL.min^{-1}) and the temperature was maintained at 20 °C using a thermostatised bath. When the photocatalytic test was carried out only with the supports, they were not activated with heating and a flow of H2 before the photocatalytic test, as oxygen reduction of TiO2 only occurs at temperatures above 460 °C, outside the working temperature range of this research (SIKHWIVHILU et al., 2007). For the catalyst with the best results, tests were carried out to reuse the catalyst, vary the radiation source with the Philips Actinic BL TL 8W G5 lamp which emits UV-A radiation (360 - 410 and 440 nm) and influence the concentration of formic acid between 5 - 30 mmol.L^{-1} .

After each experiment, the reagent solution containing the catalyst was filtered through a nylon membrane with a pore size of 0.45 μm. The filtrate was kept to determine possible leached metals and the retained catalyst was dried in an oven at 105 °C for 12 h for future analyses and reuse.

1.4 DETERMINATION OF NITRATE CONVERSION, PHOTOCATALYTIC ACTIVITY AND SELECTIVITY TO REACTION PRODUCTS.

During the tests, 2 mL aliquots were taken from the solution in the reactor at times -30, 0, 5, 30, 60, 120, 180, 240, 300 and 360 minutes to determine NO_2^-, NO_3^- and NH_4^+ to calculate the activity of the catalyst (A), the percentage of nitrate conversion ($C_{NO_3^-}$) and the selectivities to NO_2^- ($S_{NO_2^-}$), NH_4^+ ($S_{NH_4^+}$) and N2 (S_{N_2}), as shown in Figure 3. For the calculations, it was assumed that no other NOx species were formed during the photocatalytic tests (ZHANG et al., 2005), The -30 min aliquot corresponded to the collection of the reagent solution before placing it in the photocatalytic reactor containing the catalyst. After adding the nitrate solution to the reactor, it remained in contact with the catalyst for 30 min under a flow of nitrogen to assess whether the effect of nitrate adsorption was significant for the catalyst used in the test. The 0 min aliquot was collected immediately after the lamp was switched on in the reactor to start the photocatalytic reaction. The difference in nitrate concentration between these two aliquots allowed us to infer how much of the analyte had been adsorbed by the catalyst.

$$C_{NO_3^-}(\%) = \frac{[NO_3^-]_0 - [NO_3^-]_t}{[NO_3^-]_0} \times 100$$

$$A\ (\mu mol\,NO_3^-.min^{-1}.g_{cat}^{-1}) = \frac{(n^\circ\ de\ moles\ de\ NO_3^-) \times 10^{-6}}{(tempo\ observado) \times (massa\ de\ catalisador)}$$

$$S_{NO_2^-}(\%) = \frac{[NO_2^-]_t}{[NO_3^-]_0 - [NO_3^-]_t} \times 100 \qquad S_{NH_4^-}(\%) = \frac{[NH_4^+]_t}{[NO_3^-]_0 - [NO_3^-]_t} \times 100$$

$$S_{N_2}(\%) = 100 - S_{NO_2^-}(\%) - S_{NH_4^-}(\%)$$

Figura 3. Equations for nitrate conversion, photocatalytic activity and selectivity to N2, NO_2^- and NH_4^+, where $[NO_3^-]_0$ is the initial nitrate concentration and $[NO_3^-]_t$, $[NO_2^-]_t$ and $[NH_4^+]_t$ are the nitrate, nitrite and ammonium concentrations at a time t (BARBOSA et al., 2013).

1.5 DETERMINATION OF NITRATE, NITRITE AND AMMONIUM.

The aliquots collected from the solution contained in the photocatalytic reactor were filtered through nylon membranes with a pore size of 0.45 |im to remove the catalyst from the solution. Nitrate and nitrite were determined by High Performance Liquid Chromatography (HPLC) using Young Lin YL9100HPLC equipment using a C18 250x4.6x0.005 mm chromatographic column, UV/VIS detector with a wavelength of 210 nm, according to the methodology described by Epron et al. (2001). The ammonium ion was determined by UV-VIS molecular absorption spectroscopy using the Femto 800 XI equipment, using the indophenol method (APHA et al., 2012). These analytical determinations were carried out in the IFG chemistry laboratories.

1.6 CHARACTERISATION OF THE CATALYSTS.

For the determination of silver, the catalysts were digested in 15 mL of a 9 mol.L^{-1} HCl solution in open reflux at 90 °C until the solution was almost dry, which was completed to 25 mL with the 9 mol.L^{-1} HCl solution to be analysed. For the determination of palladium, copper and tin, the digestion was carried out in aqua regia solution at closed reflux at 80 °C for 3 hours and then the volume was topped up to 25 mL with water. The concentration of copper and silver in the catalysts and in the filtered solution at the end of the reaction was determined by atomic absorption spectroscopy using UNICAM 939 AA equipment, palladium was quantified using Perkin Elmer AAnalyst 200 equipment and tin using Perkin Elmer ICP-OES 7300DV. Scanning electron microscopy images were taken of solid samples deposited on carbon tape over an aluminium sample holder and coated with gold using JEOL JSM-6610 and JEOL JSM-IT300 equipment. X-ray fractograms were obtained using a Bruker D8 Discover diffractometer using monochromatic radiation from a tube with a copper anode coupled to a *Johansson* monochromator for Ka1 operating at 40 kV and 40 mA.

2. RESULTS AND DISCUSSION

Two experimental tests were carried out without the presence of a catalyst: one with the lamp switched off and the temperature controlled, and the other with the lamp switched on and the system temperature not controlled. In both cases there was no conversion of NO_3^- , which shows that UV radiation, an increase in temperature (up to

35 °C) and the presence of formic acid in the reaction system are not enough to observe nitrate degradation, showing that if there was a reduction in nitrate in the photocatalytic test, it was due to the presence of an active catalyst in the reaction.

2.1 AS A RESULT OF THE PHOTOCATALYTIC ACTIVITY AND SELECTIVITY OF THE SUPPORTS.

Figure 4 shows the conversion of c_{NO} nitrate$_3^-$ (%) and the formation of nitrite and ammonium obtained by the catalyst supports used in the work up to a reaction time of 360 min.

It was observed that the Nfios support had little photocatalytic activity, since nitrate conversion was less than 10% and stabilised after 30 min (Fig. 4A). On the other hand, NCTiO2 and P25 showed nitrate conversions of over 90% at the end of the reaction and were very similar to each other throughout the period analysed and, consequently, photocatalytic activity was close (Fig. 4A). These last two supports managed to reduce nitrate to a concentration below 50 mg NO_3^- .L^{-1} , the value recommended by the WHO, in approximately 150 min (WHO, 2011).

Virtually no nitrite was produced by the supports analysed during the photocatalytic reaction (Fig. 4B).

All the supports produced ammonium from the start of the photocatalytic reduction. P25 produced the most, with approximately 7.5 mg NH_4^+ .L^{-1} detected at the end of the reaction, around 2 mg NH_4^+ .L^{-1} more than was observed for NCTiO2. Nfios produced less than 1 mg NH_4^+ .L^{-1} (Fig. 4C).

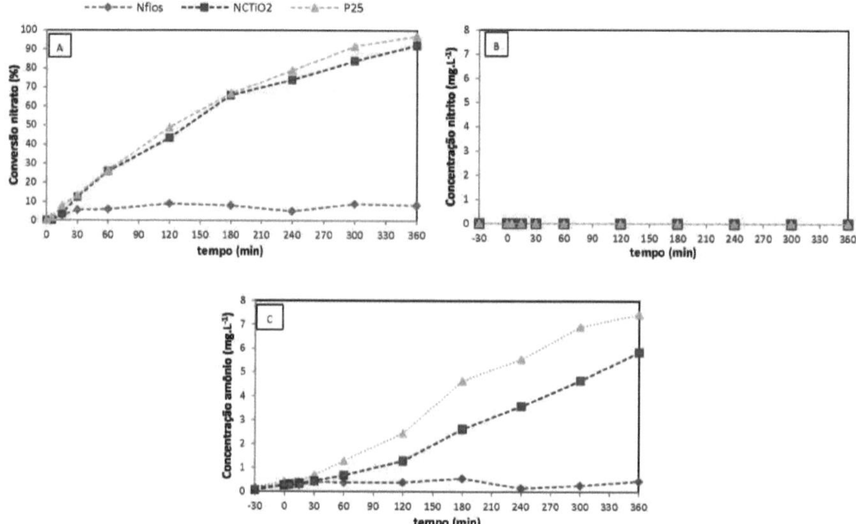

Graph of the conversion of nitrate (Fig. A), formation of nitrite (Fig. B) and ammonium (Fig. C) by the Nfios, NCTiO2 and P25 supports with a reaction time of 360 min.
[HCOOH] = 10 mmol.L^{-1} and solution temperature 20 °C.

The results obtained with P25 and NCTiO2 are surprising due to the high conversion of NO_3^- and low formation of NH_4^+ (7.4 and 5.8 mg NH_4^+.L^{-1} , respectively), as there are no reports in the literature of this combination. Sá et al. (2009) converted 100% of NO_3^- in 3 hours of reaction using P25 in photocatalytic reduction, but with selectivity to N2 of 58% and formation of just under 42 mg NH_4^+.L^{-1} . This difference to that obtained in this work can be explained by the reaction temperature used by the authors (34°C), which increases photocatalytic activity but decreases selectivity to N2 according to them. Yang et al. (2013) using Evonik P90 TiO2, which has 88% anatase phase, also achieved 100% conversion of NO_3^- , but with NH_4^+ selectivity of 17%, higher than the 5 and 8% obtained by the P25 and NCTiO2 supports tested. On the other hand, Zhang et al. (2005) and Gao et al. (2004) obtained 15.2% and 0% nitrate conversion with P25, respectively, much lower than that reported in the above works and in this study.

The conversion obtained, even without activating the support with H2 and heating, can be explained by the photoconductance of TiO2 in the P25 and NCTiO2 supports. The reduction of nitrate can occur both by the electrons photogenerated on the supports and by the H2 present in the aqueous medium from the degradation of formic acid (SOARES et al., 2014). Surface reduction can be provided by the creation of exposed Lewis acid sites (oxygen vacancies) that adsorb and reduce nitrate and/or nitrite on Ti centres^{3+} (SÁ et al., 2005; SOARES et al., 2014). In a review article, Fujishima et al. (2008) proposed that photocatalysis on the surface of TiO2 occurs on two different types of Ti-OH groups, which play important roles in trapping charges, Ti4+-*OH radicals (gap trappers) and Ti^{3+} -OH groups (electron trappers), which explains the activity of the supports. Sá et al. (2005) believe that the nitrate reduction process can also take place at the boundaries or edges between the anatase and rutile crystalline structures, which are present in the P25 and NCTiO2 supports.

The fact that no nitrite was formed in the experiments with the supports suggests that the reduction of nitrate to ammonium and molecular nitrogen occurs at the same active site on the surface of the supports, as described by Sá et al. (2005). Table 1 illustrates the similarity between the results of P25 and NCTiO2 in terms of product selectivity when 75% of NO3 was converted^{-} . However, P25 converted 75% of NO3^{-} more quickly than NCTiO2, around 26 minutes earlier, thus resulting in greater photocatalytic activity. The difference in the conversion of NO3^{-} between P25 and NCTiO2 at the end of the reaction may in part be due to the effect of nitrate adsorption (difference in NO3^{-} concentration between the -30 and 0 min aliquots), which in the former support was around 5% of the total contained in the solution compared to 2% in the latter, results not shown in the work.

Table 1. Experimental data of the supports regarding activity and selectivity to nitrite, ammonium and nitrogen, and reaction time when nitrate conversion is 75%.

Brackets	75% conversion of NO3^{-}				
	Activity (gmolNO3-.min-1.gcat)$^{-1}$	$S_{NO_2^-}$ (%)	$S_{NH_4^+}$ (%)	S_{N2} (%)	Time (min)
P25	8,5	0	7	93	222
NCTíO2	7,5	0	7	93	248

2.2 RESULTS OF THE PHOTOCATALYTIC ACTIVITY AND SELECTIVITY OF MONOMETALLIC CATALYSTS.

The monometallic Pd catalysts in proportions of 1 and 5% in relation to the total mass of catalyst (mt) had photocatalytic nitrate conversion below 10% and, consequently, very low activity, similar to that found by Gao et al. (2004) and Soares et al. (2014). However, some studies on the catalytic reduction of nitrate in the presence of H2 have reported high activity and conversion for these catalysts supported on P25 (BARRABÉS & SÁ, 2011; SÁ et al., 2005).

The monometallic Ag catalysts produced had NO_3 conversion$^-$ greater than 40% in only two cases, with 1% (mt) supported on P25 and NCTiO2.Figure 5 shows the NO_3 conversion$^-$ and the formation of NO_2^- and NH_4^+ during the 360 min reaction period for the P25-Ag1% and NCTiO2-Ag1% catalysts.

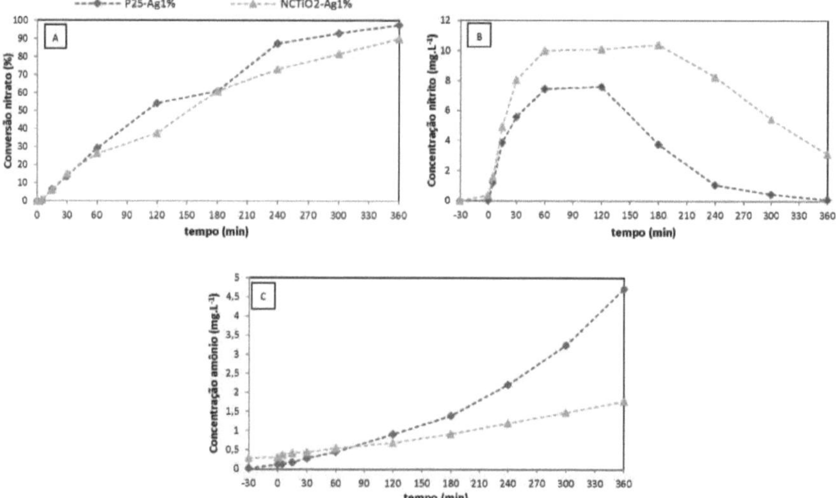

Figura 5.   Graph of nitrate conversion (Fig. A), nitrite formation (Fig. B) and ammonium (Fig. C) in the presence of the monometallic catalysts P25-Ag1% and NCTiO2-Ag1%, with a reaction time of 360 min. [HCOOH] = 10 mmol.L-1 and solution temperature 20 °C.

In terms of nitrate conversion, P25-Ag1% stood out with 98% compared to 90% for NCTiO2-Ag1% at the end of the reaction. However, they had very similar conversion curves, especially at the start of the reaction, which diverged after 60 min and only intersected again at 180 min, when 60% conversion had already occurred.

There was significant nitrite formation in the tests with both catalysts, with a maximum of around 10 mg.L^{-1} with NCTiO2-Ag1% in 180 min. At the end of the reaction, only P25-Ag1% managed to reduce all the nitrite formed previously. On the other hand, NCTiO2-Ag1% did not degrade this analyte within the six hours of the experiment, remaining 3 mg.L^{-1} at the end of the experiment. The nitrite formation curves showed that nitrite is an intermediate in the photocatalytic nitrate reduction reaction, since it is formed in considerable quantities and is degraded over time. In this way, it can be seen that, unlike what was observed with the supports, there was a predominance of the two-stage reaction route, the first consisting of the reduction of nitrate with the formation and subsequent desorption of nitrite and the second the reduction of nitrite to molecular nitrogen and/or ammonium.

The formation of ammonia by P25-Ag1% was more than three times that of NCTiO2-Ag1%, around 5 mg.L^{-1} compared to 1.5 mg.L^{-1} , maintaining the trend seen with the supports in which P25 produced more ammonium than NCTiO2.

Comparing the performance of the supports and catalysts, it can be inferred that metal deposition led to the formation of nitrite adsorbing/reducing active sites very close to each other, which led to less ammonium being formed, since the generation of ammonium is disadvantaged by the proximity of sites that adsorb and reduce nitrite (PRUSSE & VORLOP, 2001).

The N2 selectivity data (Table 2) showed that the two catalysts were very similar, although P25-Ag1% degraded 75% of the nitrate 42 minutes earlier than NCTiO2-Ag1%, showing greater photocatalytic activity. When comparing the selectivity of the reaction products when 75% nitrate was converted, NCTiO2-Ag1% had 6% higher selectivity for NO2^{-} and 5% lower selectivity for NH4^{+} than P25-Ag1%. The active nitrite reduction sites on the NCTiO2-Ag1% catalyst possibly have a closer interaction with each other and are more aggregated than those on P25-Ag1%, due to the lower formation of ammonium and difficulty in degrading the intermediate. The aggregation of these sites hinders the rapid adsorption/reduction of nitrite, as the anions bound to the surface repel those dispersed in the solution, preventing access and leading to the saturation of these sites for a longer period.

Table 2. Experimental data for the P25-Ag1% and NCTiO2-Ag1% catalysts in terms of activity and selectivity for nitrite, ammonium and nitrogen, with a reaction time corresponding to 75% nitrate conversion.

Catalyst	75% conversion of NO_3^-				
	Activity ($\mu mol NO_3^-$ min^{-1}.gcat)$^{-1}$	S_{NO2^-}(%)	S_{NH4^+}(%)	S_{N2} (%)	Time (min)
P25-Ag1%	7,8	4	7	89	212
NCTiO2-Ag1%	7,5	10	2	88	254

The silver monometallic catalysts supported on P25 had different results to those achieved by Zhang et al. (2005), who obtained 98% nitrate conversion with 100% N_2 selectivity in 30 minutes. The two catalysts showed high resistance to leaching in the photocatalytic reaction carried out, since no Ag was detected in the filtered solution after the reaction and there were no significant differences in Ag between the new and used catalysts, as shown in Table 3.

Table 3. Percentage of Ag on the surface of P25 and NCTíO2 supports, in new and once-used catalysts.

Sample	% (m/m) of metal in new catalyst	% (m/m) of metal in used catalyst
P25-Ag1%	0,9	0,8
NCTiO2-Ag1%	1,0	1,2

2.3 RESULTS OF THE PHOTOCATALYTIC ACTIVITY AND SELECTIVITY OF BIMETALLIC CATALYSTS.

Two metals (Cu and Sn) whose function is to promote nitrate reduction were selected in an attempt to increase the efficiency of the catalysts compared to the supports and monometallic catalysts (BARBOSA et al., 2013; SOARES et al., 2014; DODOU- CHE et al., 2009). Thus, the deposition of a second metal on the surface of the monometallic catalyst can lead to a decrease in the electron/laccase recombination rate of the semiconductor due to better charge separation in the material, making the photocatalytic nitrate reduction reaction faster and easier (SÁ et al., 2009).

Only two bimetallic catalysts supported on P25-Ag5% were produced, one with 2% copper (Cu) and the other with 2% tin (Sn), which did not show photocatalytic conversion.

The bimetallic catalysts supported on Nfios-Ag5% with Cu or Sn ratios of 0.5, 2 and 4% (mt) did not show photocatalytic nitrate conversion of more than 15% after

360 min of reaction. Only 2 catalysts supported on Nfios-Pd5% were synthesised, since this monometallic catalyst showed very low photocatalytic activity. The Nfios-Pd5% with 2% Cu did not convert nitrate, while the one with 2% Sn managed to convert 20%.

Figure 6 shows the results of the bimetallic catalysts in terms of nitrate conversion, nitrite and ammonium formation only for those that had photocatalytic nitrate conversion above 50%, all of which were supported on NCTiO2.

As the monometallic NCTiO2-Ag1% catalyst had a conversion of over 90%, it was decided to use it as a support for a second metal (Cu or Sn). In this case, two catalysts stood out with photocatalytic conversion greater than 58% after 360 min of reaction, NCTiO2-Ag1%-Sn0.5% and NCTiO2-Ag1%-Cu2% (Fig. 6). Using NCTiO2-Ag5% as a support, conversions of more than 20% were not achieved. The deposition of a second metal on monometallic Ag catalysts in any proportion did not improve the material's performance in terms of its photocatalytic activity and nitrate conversion.

Although the NCTiO2-Pd5% catalyst showed negligible photocatalytic activity, the second metal (Cu or Sn) was deposited on it, since bimetallic catalysts with these metals are described in the literature as being very active for nitrate reduction (BARBOSA et al., 2013; SOARES et al., 2014; PRUSSE & VORLOP, 2001; DO-DOUCHE et al, 2009; KOMINAMI et al., 2005; PALOMARES et al., 2010; SOARES et al., 2011; ZHANG et al., 2008; YANG et al., 2011; KIM et al., 2013; GAO et al., 2003; BAR- RABÉS et al., 2006; JUNG et al., 2012; SOARES et al., 2012; SOARES et al., 2010; GAR- RON et al., 2005; GUO et al., 2012; HIRAYAMA et al., 2014; WEHBE et al., 2009; WANG et al., 2009).

Surprisingly, the supported catalysts containing Pd-Cu, so well described in the literature, did not achieve nitrate degradation values above 30 per cent. NCTiO2- Pd5%-Cu0.5% had the best result, converting 28% of the initial nitrate. Soares et al. (2014) achieved photocatalytic conversion of over 80% using 80 mmol.L^{-1} of formic acid as the electron donor agent for the reaction. Considering catalytic reduction with H2 flow during the reaction, bimetallic Pd-Cu catalysts supported on TiO2 have achieved nitrate conversion close to or equal to 100% in various works (GAO et al., 2003; KIM et al., 2013; SOARES et al., 2011; WEHBE et al., 2009; YANG et al., 2011; ZHANG et al., 2008).

The bimetallic Pd and Sn catalysts have been identified as having the greatest activity and selectivity in the catalytic reduction of nitrate using H2 during the reaction, although no references were found to studies using them in the photocatalytic reduction of nitrate (BARRABÉS & SÁ, 2011).

All the Pd and Sn bimetallic catalysts supported on NCTiO2 tested had improved performance compared to the support and monometallic catalysts studied. Conversion was greater than 90 per cent after 360 minutes. The one with Sn at 2% (mt) had the highest photocatalytic activity, reaching over 99% nitrate conversion in 240 minutes, among all the Pd and Sn catalysts tested.

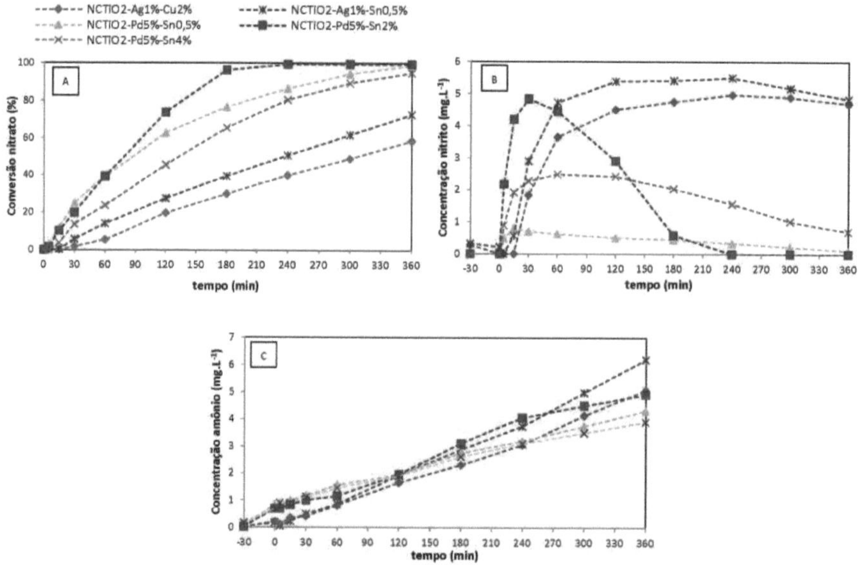

Figura 6. Graph of the conversion of nitrate (Fig. A), formation of nitrite (Fig. B) and ammonium (Fig.
C) by the bimetallic catalysts NCTiO2-Ag1%-Cu2%, NCTiO2-Ag1%-Sn0.5%, NCTiO2- Pd5%-Sn0.5%, NCTiO2-Pd5%-Sn2% and NCTiO2-Pd5%-Sn4% with a reaction time of 360 min. [HCOOH] = 10 mmol.L^{-1} and solution temperature 20 °C.

The NCTiO2-Ag1%-Cu2% and NCTiO2-Ag1%-Sn0.5% catalysts showed almost linear nitrate conversion during the 6-hour reaction. The formation of nitrite was very pronounced with a maximum at approximately 4.5 and 5.5 mg.L^{-1} , respectively. The evolution of nitrite shows a downward trend at the end of the 6-hour reaction, which suggests that the reaction is not close to equilibrium and that, over time, more nitrite will have to be converted to other nitrogen species. These catalysts also form the most ammonium of those shown in Figure 6, around 5 and 6 mg.L^{-1} , respectively.

The formation of nitrite for the bimetallic Pd and Sn catalysts is similar in that there is a maximum concentration between 15 and 60 min, with a subsequent drop as the reaction progresses. The highest concentration of nitrite among these catalysts is observed with NCTiO2-Pd5%-Sn2%, which produces 5 mg.L^{-1} in 30 min, but is the only one that completely degrades it in 240 min of reaction. The other two (NCTiO2-Pd5%-Sn4% and NCTiO2-Pd5%-Sn0.5%) leave nitrite residues of approximately 0.7 and 0.1 mg.L^{-1} , respectively, at the end of the reaction period. The formation of ammonium for these catalysts is very close, at between 4 and 5 mg.L^{-1} at the end of the experiment.

No studies were found in the literature with bimetallic Pd and Sn catalysts in any proportions supported on TiO2 used for the photocatalytic or catalytic reduction of nitrate for comparison purposes. However, several studies have used alumina as a support for these metals to study the catalytic reduction of nitrate, which have obtained great activity and selectivity for N2 (PRUSSE & VORLOP, 2001; PALOMARES et al., 2010; GARRON et al., 2005). Other supports have also been reported with varying proportions of Pd and Sn, obtaining great activity and selectivity to N2, such as silica (GARRON et al., 2005), sulphonated poly(styrene-divilbenzene) (BARBOSA et al., 2013), polypyrrole and poly(nylline) (DODOUCHE et al., 2009).

The NCTiO2-Pd5%-Sn2% catalyst showed the greatest photocatalytic activity when converting 75% of the initial nitrate, doing so 49 and 95 minutes before the NCTiO2-Pd5%-Sn0.5% and NCTiO2-Pd5%-Sn4% catalysts, respectively (Table 4). Although the selectivity of the products obtained by NCTiO2-Pd5%-Sn2% was only better in relation to NH4^{+} , the selectivity for NO2^{-} and N2 was very close to that found for the NCTiO2-Pd5%-Sn0.5% and NCTiO2-Pd5%-Sn4% catalysts (Tab.4). The amount of NO2^{-} produced by NCTiO2-Pd5%-Sn2% in 124 min (75% nitrate conversion) was 2.8 mg.L^{-1} below the values recommended by the WHO (WHO, 2011).

Differences in activity and selectivity to N2 between bimetallic catalysts with

Pd and Sn, monometallic catalysts and supports can be understood by the change in the distribution of active sites in the reduction of nitrate and nitrite (PRUSSE & VORLOP, 2001). It is suggested that in these bimetallic catalysts the higher activity and lower nitrite formation are due to the greater number of active sites that reduce nitrate and nitrite. The greater formation of ammonium, on the other hand, is due to the greater dispersion of nitrite reduction sites, which favours the generation of ammonium.

Table 4. Experimental data for the NCTiO2-Pd5%-Sn0.5%, NCTiO2-Pd5%-Sn2% and NCTiO2-Pd5%-Sn4% catalysts in terms of activity and selectivity for nitrite, ammonium and nitrogen, and reaction time when nitrate conversion is 75%.

Catalyst	75% conversion of NO_3^-				
	Activity (gmolNO_3^-.min^{-1}.gcat)$^{-1}$	$S_{NO_2^-}$ (%)	$S_{NH_4^+}$ (%)	S_{N_2} (%)	Time (min)
NCTiO2-Pd5%-Sn0.5%	9,0	1	4	95	173
NCTiO2-Pd5%-Sn2%	12,4	5	3	92	124
NCTiO2-Pd5%-Sn4%	6,5	3	5	92	219

The bimetallic catalysts with silver, Table 5, leached practically all of the second deposited metal, copper and tin, after the photocatalytic test. The bimetallic catalysts with Pd, on the other hand, showed no signs of leaching of either of the two deposited metals.

Table 5. Percentage of metals added and deposited on the surface of the NCTiO2 support, for new and once-used catalysts.

Sample	% (m/m) of metal in new catalyst		% (m/m) of metal in used catalyst	
	1st metal	2nd metal	1st metal	2nd metal
NCTiO2-Ag1%-Cu2%	0,7	1,8	0,7	0,1
NCTiO2-Ag1%-Sn0.5%	1,0	0,2	0,7	0
NCTiO2-Pd5%-Sn0.5%	5,8	0,15	6,2	0,15
NCTiO2-Pd5%-Sn2%	5,9	1,3	5,9	1,3*
NCTiO2-Pd5%-Sn4%	5,4	1,2	5,5	1,2

Digestion carried out with aqua regia in closed reflux at 80°C for 3 hours. *After 5 photocatalytic tests.

It can be seen that the amount of Sn added is not completely deposited on the surface of the monometallic catalyst and the addition of more than 2% Sn (mt) did not increase the amount impregnated on the surface, which shows that the Sn was deposited on the surface of the monometallic catalyst at specific sites that became saturated, as can be seen in table 5.

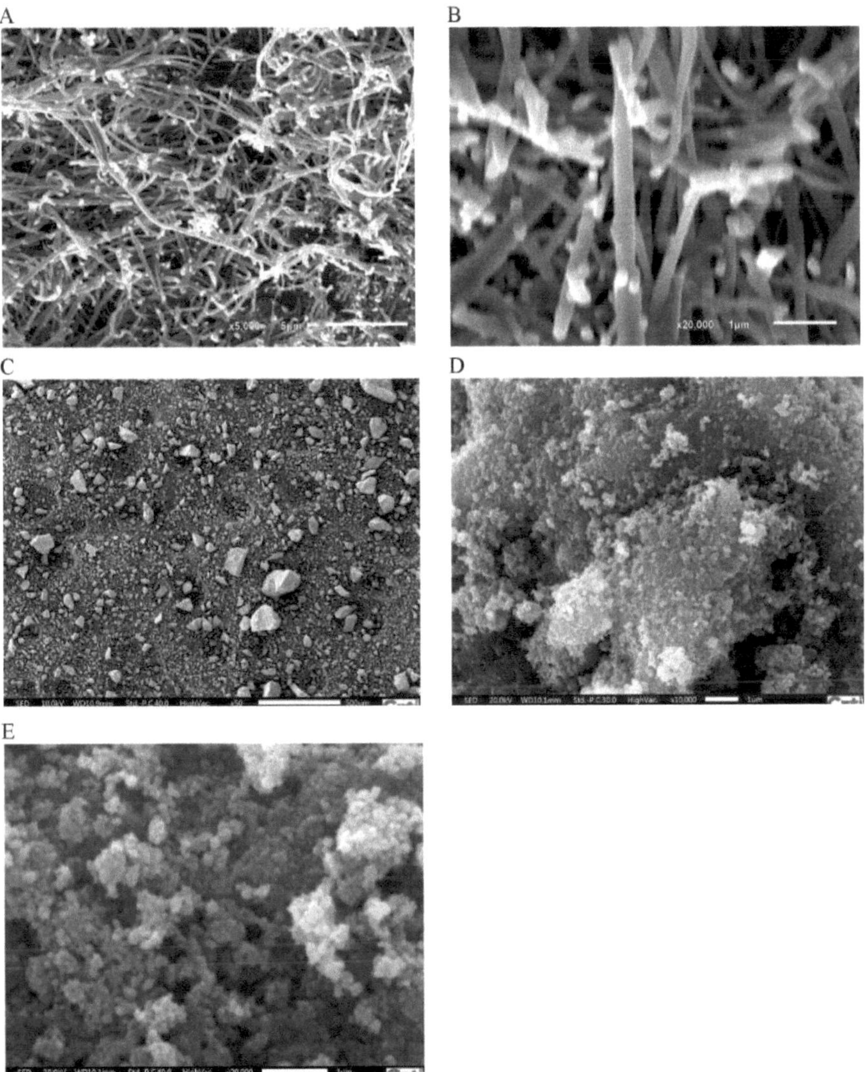

Figura 7. Scanning electron microscopy (SEM) images of the nanowires (A and B) and the NCTiO2-Pd5%-Sn2% catalyst (C, D and E).

Although there was a similar amount of Sn in the NCTiO2-Pd5%-Sn2% and NCTiO2-Pd5%-Sn4% catalysts, the photocatalytic tests of these two materials gave quite different results (Tab. 5). This is possibly explained by the different distribution of Sn on the surface of the two catalysts, due to the different concentrations of Sn solutions initially added to the monometallic catalyst.

Figure 7 shows the SEM images of the nanowires and the NCTiO2-Pd5%-Sn2% catalyst. The nanowire diameters are on average 250 to 300 nm and lengths of tens of micrometres, which can reach the millimetre scale, since it is not possible to see the beginning and end of the material (Fig. 7A and B). The NCTiO2- Pd5%-Sn2% catalyst has granules of very different sizes, from around 250 |im to less than 1 |im (Fig. 7C). As you zoom in on the image (Fig. 7D and E), you can see that these heterogeneously distributed granules are clusters of small particles with an average diameter of approximately 100 nm, forming pores and a very rough surface. Figures 7B and 7E show at 20,000 magnification that the nanowires have less surface area and roughness than the NCTiO2-Pd5%-Sn2%, which partly explains the nanowires' lower photocatalytic activity. Since there is greater contact between the surface of NCTiO2-Pd5%-Sn2% and nitrate, and a greater number of edges on the surface of NCTiO2- Pd5%-Sn2%, this leads to an increase in active nitrate reduction sites.

2.4 NCTiO2-Pd5%-Sn2% CATALYST

Based on the results presented, some reaction variables were optimised using NCTiO2-Pd5%-Sn2%, which had the highest photocatalytic activity. The effects of electron donor concentration, catalyst activation, radiation source and the physicochemical stability of the catalyst after being reused five times were observed.

2.4.1 EFFECT OF VARYING THE CONCENTRATION OF THE ELECTRON DONOR (FORMIC ACID)

Short-chain carboxylic acid molecules are widely used in photocatalytic nitrate reduction reactions because they lead to high conversions and because they are common products in the incomplete mineralisation of organic compounds that have undergone advanced oxidative treatment (AOP), which facilitates a possible future application combining oxidative and reductive treatments of contaminated water (SHAND & ANDERSON, 2013). Formic acid has been used as a model electron donor in photocatalytic nitrate reduction studies and its use is justified by the high conversion observed when it is present in the solution compared to other reducers, in addition to the fact that the products of its decomposition buffer the reaction medium, inhibiting the formation of ammonia (SÁ et al., 2009).

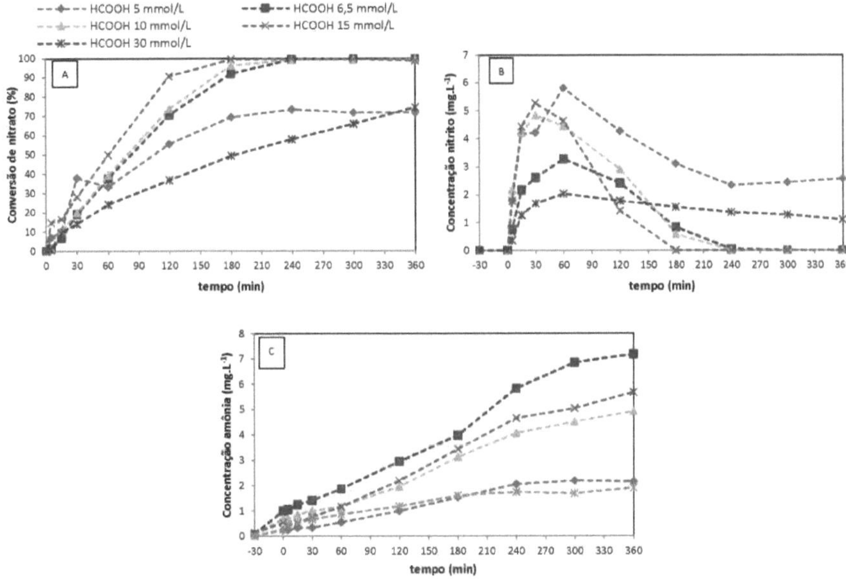

Figura 8. Graph of nitrate conversion (Fig. A), nitrite formation (Fig. B) and ammonium (Fig. C) at different concentrations (5 mmol.L^{-1} ; 6.5 mmol.L^{-1} ; 10 mmol.L^{-1} 1; 15 mmol.L^{-1} and 30 mmol.L^{-1}) of formic acid (HCOOH) up to a reaction time of 360 min. Reaction temperature 20°C.

Figure 8 shows the nitrate conversion, nitrite and ammonium formation curves for different concentrations of formic acid (HCOOH) from 5 to 30 mmol.L^{-1} using NCTiO2-Pd5%-Sn2%. Fig. 8A shows that photocatalytic conversion depends on the concentration of HCOOH, since at concentrations between 6.5 and 15 mmol.L^{-1} there was 100% conversion in 240 min. When 5 mmol.L^{-1} of HCOOH was used in the reagent solution, 72% conversion was achieved in 240 min, and this level was maintained until the end of the reaction, indicating stabilisation of the photocatalytic reduction due to consumption of all the reducing agent. This showed that 5 mmol.L^{-1} of HCOOH is insufficient for the complete degradation of nitrate, which is essential for this reaction to take place. Similar conversion was obtained with 30 mmol.L^{-1} , but the evolution of nitrate conversion over time was practically linear, indicating that the excess HCOOH may strongly adsorb to the catalyst surface, preventing nitrate from accessing the active sites responsible for its reduction (FUJISHIMA et al., 2008; SOARES et al., 2014; ZHANG et al., 2005). Thus, competition for active sites reduces the speed of nitrate reduction and its complete conversion. This behaviour, in the presence of a high concentration of reductant in the solution, suggests that the nitrate and nitrite reduction reaction occurs preferentially on the active sites on the catalyst surface, rather than in reactions with radical species (CO_2^-) generated by the catalyst.

from the degradation of HCOOH, as proposed in some articles (SHAND & AN-DERSON, 2013; ZHANG et al., 2005).

The nitrite formation curves (Fig. 8B) are generally very similar to those presented in this work. However, when 5 mmol.L^{-1} of HCOOH was present, from 240 min onwards, nitrite degradation stabilised, at the same time as nitrate conversion also ceased, corroborating the explanation that the amount of HCOOH is insufficient for the reaction to proceed. The nitrite evolution curve, considering 30 mmol.L^{-1}, also differed from the other curves, as there was a maximum at 60 min with a slight drop until 360 min, which showed that nitrite also competes with HCOOH, which is strongly adsorbed, for the active sites on the surface of the material (FUJISHIMA et al., 2008).

The experiment that showed the highest concentration of ammonium (7.2 mg.L^{-1}) at the end of the reaction was the one with 6.5 mmol.L^{-1} of HCOOH, which may be related to the higher pH of the reaction medium (Table 6) in relation to those experiments that converted more than 90% of nitrate, as this increase favours the formation of ammonium (SHAND & ANDERSON, 2013). The lowest ammonium concentrations were observed in the experiments with 5 and 30 mmol.L^{-1} of HCOOH, due to the lower nitrate conversion obtained. The tests with 6.5, 10 and 15 mmol.L^{-1} of reductant formed ammonium even after the total degradation of nitrate and nitrite in 240 min, which shows that residual nitrogenous species adsorbed on the catalyst preferentially undergo *overreduction*.

Table 6 illustrates the variation in pH for the different concentrations of HCOOH in the solution, which is explained by the ion exchange between the hydroxyls (OH^-) of titanols (Ti-OH) and nitrate and/or nitrite, which causes the former to be released into the solution and maintain the electroneutrality of the solution [11], and also by the consumption of hydronium (H^+) in the photocatalytic reduction reactions [9,12]. The decrease in the amount of HCOOH showed the limits of this acid's buffering capacity in the tests. The pH of the solution at the end of the reactions with 5 and 6.5 $mmol.L^{-1}$ indicated the minimum amount of reductant that can be added to achieve 100% nitrate conversion with the catalyst studied. On the other hand, the presence of excess reductant compromises the speed of the photocatalytic reaction due to the difficulty of nitrate and/or nitrite accessing the adsorption/reduction sites on the catalyst surface and there was no pH variation at the end of the experiment. Therefore, the ideal concentration of formic acid was between 10 and 15 $mmol.L^{-1}$, which in 180 min of reaction degraded all the initial nitrate, there was a very small amount of nitrite, as well as ammonium formation below 6 $mg.L^{-1}$.

Table 6. The influence of HCOOH concentration on the conversion of nitrate $C_{NO_3^-}$ (%) and formation of NO_2^- and NH_4+ by NCTiO2-Pd5%-Sn2%.

Conc. HCOOH $mmol.L^{-1}$	$C_{NO_3^-}$ (%)	NO_2^- formed $(mg.L)^{-1}$	NH_4+ formed $(mg.L)^{-1}$	pH*
5	69	3,1	1,5	3,3 - 8,1
6,5	92	0,8	4,0	3,0 - 4,6
10	96	0,6	3,1	3,0 - 3,9
15	99	0	3,4	2,9 - 3,3
30	49	1,5	1,6	2,9 - 2,9

Reaction conditions: initial concentration of 100 mg $NO_3^-.L^{-1}$, solution at 20 °C and irradiation time of 180 min. *In 360 min of irradiation.

In this work, 100% nitrate conversion was achieved with small amounts of electron donor, compared to what has been reported in the literature, which is advantageous due to the smaller amount of residual HCOOH or $HCOO^-$ at the end of nitrate degradation (SÁ et al., 2009; SHAND & ANDERSON, 2013; SOARES et al., 2014; WANG et al., 2009; WEHBE et al., 2009; YANG et al., 2013; ZHANG et al., 2005).

An important parameter to be analysed is the molar ratio between HCOOH and NO_3^-, in which variations theoretically determine which products are favoured, for the degradation of nitrate to N2 this ratio should be 2.5, for greater formation of nitrite and ammonium the values should be 1 and 4, respectively (YANG et al., 2013), as shown in Figure 9.

$$NO_3^- + HCOOH \rightarrow NO_2^- + CO_2 + H_2O, \tag{1}$$

$$2NO_3^- + 5HCOOH + 2H^+ \rightarrow N_2 + 5CO_2 + 6H_2O, \tag{2}$$

$$NO_3^- + 4HCOOH + 2H^+ \rightarrow NH_4^+ + 4CO_2 + 3H_2O, \tag{3}$$

Figura 9. Chemical equations for the reaction of nitrate with formic acid [15].

The optimum ratio found in this work for the NCTiO2-Pd5%-Sn2% catalyst is 9, well above the theoretical ratio, values higher or lower than this have lower activity and selectivity to N2, as shown in Figure 10. The difference between the theoretical and real ratios is due to kinetic effects in the system, such as impediments to active sites, diffusion aspects, temperature, among others. Therefore, in order to achieve the best results, one should work close to this ratio, considering the experimental apparatus and variables and the catalyst used in this work.

Adding 15 mmol.L^{-1} of HCOOH to the nitrate solution to be photocatalytically degraded improved all the variables in Table 7 compared to other concentrations of the reductant. Using the above concentration of formic acid, there was an improvement in selectivity to N2, NO2$^-$ and NH4$^+$ and 75% of nitrate was converted 17 and 36 minutes before the experiments containing 10 and 6.5 mmol.L^{-1} of HCOOH.

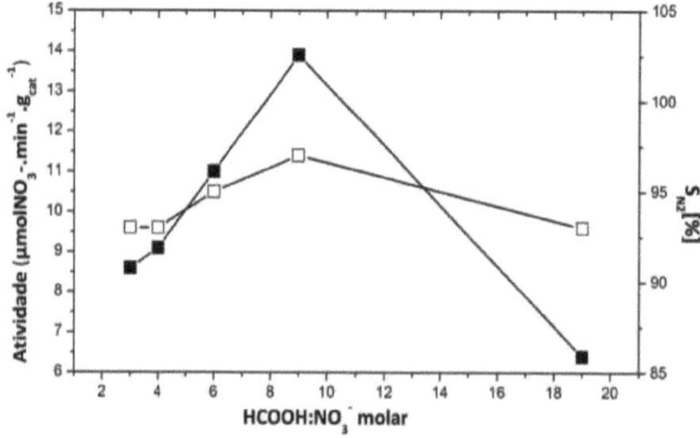

Figura 10. Graph of the molar ratio of HCOOH:NO3- by the activity and selectivity to

N2 of the NCTiO2-Pd5%-Sn2% catalyst at 180 min of irradiation. Empty square ( ) = N2 selectivity and filled square (■) = catalyst activity.

Table 7. Influence of HCOOH concentration on activity and selectivity to nitrite, ammonium and nitrogen, and on reaction time when nitrate conversion is 75% using the NCTiO2-Pd5%-Sn2% catalyst.

Conc. HCOOH mmol.L-1	75% conversion of NO_3^-				
	Activity ($gmolNO_3^-$ .m in- 1.g^{cat} 1)	SNO_2^-(%)	SNH_4^+ (%)	SN_2(%)	Time (min)
6,5	10	4	6	90	133
10	12,4	5	3	92	124
15	19,6	4	2	94	97

2.4.2 EFFECT OF CATALYST ACTIVATION AND RADIATION SOURCE

Figure 11A showed that the activation of the catalyst and the spectral range of the radiation incident on the material are decisive for the complete conversion of nitrate. The use of a UV-A lamp and the non-use of the lamp altered the nitrate conversion very little, about 9% more for the first test. The catalyst therefore needed wavelengths shorter than the UV-A region for the electronic transition process between the valence band and the conduction band to take place and for photocatalysis to take place with energetically excited electrons on the catalyst surface. When UV-C radiation was switched to, but without activating the catalyst, there was only a slight improvement in this result, indicating that the reduction reaction took place preferentially on reduced active sites on the catalyst surface generated during activation with H2 in situ. The inherent complexity of the photocatalytic process in question can therefore be seen, as it is not enough to just consider one of the variables, such as a wavelength with energy greater than the catalyst's bandgap. If there are no active sites to concentrate or trap the electrical charges generated, the conversion will not be satisfactory, and vice versa.

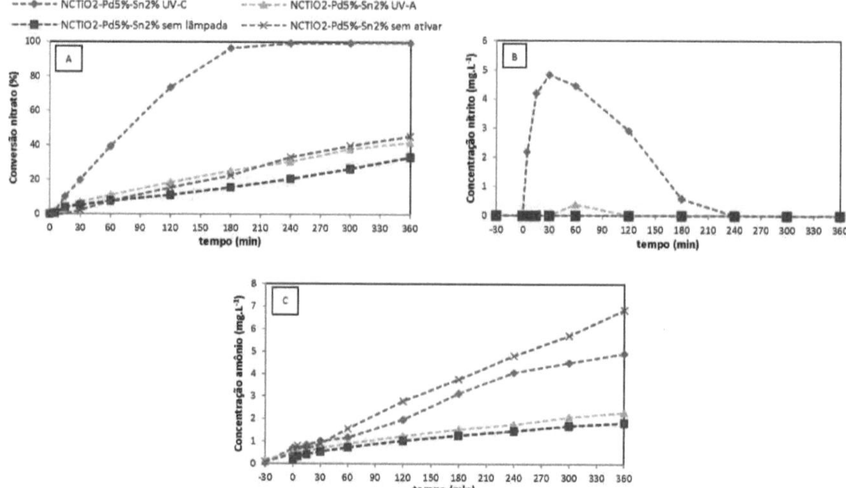

Figura 11. Graph of the conversion of nitrate (Fig. A), formation of nitrite (Fig. B) and ammonium (Fig. C) by the NCTiO2-Pd5%-Sn2% catalyst with UV-C lamp, UV-A, without lamp and without activating the catalyst and with the UV-C lamp up to the reaction time of 360 min. [HCOOH] = 10 mmol.L^{-1} and solution temperature 20 °C.

The nitrite formation curve, Figure 11B, revealed some common points in relation to the reaction mechanism that occurred with the supports. Firstly, there was practically no release of nitrite into the solution when the catalyst was not activated, the lamp was not switched on or the UV-A region was used. The reduction of nitrate in these three tests probably took place on the same active site, ultimately producing N2 or NH_4^+ , which may be the same as the precursor support, a route that may have been favoured by the lack of energy for optimal generation of electrons and gaps, or by the lack of reduced active sites that trap these charges.

What was surprising was the greater amount of ammonium produced when the catalyst was not activated, Figure 11C, which can be attributed to the favouring of the reduction of nitrite by H2 instead of the electrons trapped on the surface and/or the large dispersion of the active nitrate and nitrite adsorption sites that favoured the reduction to ammonium (BARRABÉS & SÁ, 2011; SÁ & ANDERSON, 2008).

2.4.3 EFFECT OF CATALYST REUSE

A very important aspect for the applicability of the catalyst is knowing how many times it can be reused without losing its initial activity and selectivity, resulting in a longer lifespan. Knowing whether this reuse requires activation at each experimental test is also fundamental to understanding the material's loss of efficiency. The prefix "t" and a number were used to indicate how many times the catalyst was reused, for example, t5 indicates that the catalyst was reused 5 times. When testing the catalyst that had already been used once (t1) without activating it in situ, the lowest nitrate conversion was observed among those shown in Figure 12A, showing once again that the production of reduced active sites is essential for total nitrate degradation. Now, when comparing this catalyst with the new one that was not activated, as mentioned in section 2.4.2, Figure 11A, there was a 30 per cent difference in conversion, which may be linked to the greater quantity of reduced active sites remaining from a first activation with H_2. The nitrate conversion results, shown in Figure 12A, revealed that the catalyst is not easily deactivated or poisoned, as there was a drop to 81% conversion at the end of the reaction only on the fifth time it was reused, whereas until the fourth reuse it maintained nitrate degradation above 87%.

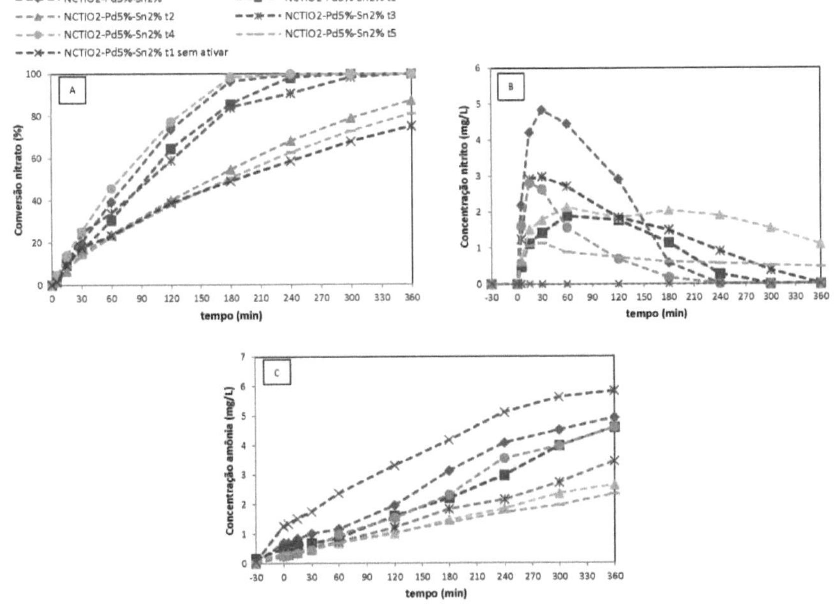

Figura 12. Graph of nitrate conversion (Fig. A), nitrite formation (Fig. B) and ammonium (Fig. C) by the NCTiO2-Pd5%-Sn2% catalyst, reused five times and reused once without activating the catalyst again up to a reaction time of 360 min.

According to the previous section, the catalyst that was not activated with H2 in situ did not release nitrite into the solution, showing that the reduction of nitrate and nitrite occurred at the same active site. It was noted that the more the catalyst is reused, the lower the maximum amount of nitrite formed and/or released into the solution. It was also observed that the degradation of nitrite during the reaction became slower in experiments t2, t3 and t5, and this phenomenon was not seen in t4. This observation indicates that in situ activation can generate reduced active sites distributed differently in each procedure, since the greater dispersion of nitrite adsorbing/reducing sites can determine the speed of its reduction and the formation of N2 or NH+ products (PRUSSE & VORLOP, 2001).

The catalyst that did not undergo activation was the one that formed the most ammonium, similar to what was seen in Section 2.4.2 Figure 11C. This finding corroborates the belief that in situ activation contributes to less dispersion and/or the creation of clusters of reduced active sites, leading to lower ammonium production. The reuse of the catalyst showed that the formation of ammonium decreased, Figure 12C, as seen for nitrite, Figure 12B, and in the test with the catalyst reused for the fifth time the lowest amount of ammonium formed was observed.Table 8 shows that after the first test there was an improvement in nitrite and nitrogen selectivity until the last test carried out, ammonium selectivity was practically maintained. There were variations in the degradation time of 75% nitrate and the photocatalytic activity, possibly due to the non-reproducibility of the in situ activation of the catalyst.

Table 8. Influence of reusing the NCTiO2-Pd5%-Sn2% catalyst on activity and selectivity to nitrite, ammonium and nitrogen, and on reaction time when nitrate conversion is 75%.

Catalyst	75% conversion of NO_3^-				
	Activity ($\mu molNO_3$ $-..min^{-1} .gcat)^{-1}$	S_{NO2^-} (%)	$S_{NH_4^+}$ (%)	S_{N2} (%)	Time (min)
NCTiO2-Pd5%-Sn2 %	12,4	5	3	92	124
NCTiO2-Pd5%-Sn2% t1	11,6	2	3	95	150
NCTiO2-Pd5%-Sn2% t2	6,7	2	3	95	279
NCTiO2-Pd5%-Sn2%t3	12,0	2	2	96	159
NCTiO2-Pd5%-Sn2% t4	15,1	1	2	97	116

NCTiO2-Pd5%-Sn2% **t5**	5,6	1	3	96	317
NCTiO2-Pd5%-Sn2% **t1 without activating**	4,5	0	9	91	360

In general, the results regarding the reuse of the catalyst were very promising, as N2 selectivity was above 95% and nitrate conversion was only not complete in the second and fifth tests, but with conversion above 81%. Even when not activated, the t1 catalyst produced high N2 selectivity and considerable nitrate degradation (75% nitrate conversion).

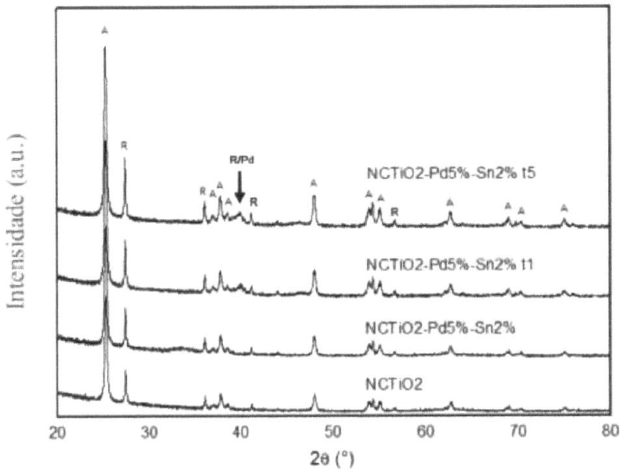

Figura 13. X-ray diffractograms of the new NCTiO2, NCTiO2-Pd5%-Sn2% support, tested
once and five times. A = *anatase; R = rutile. R/Pd = rutile or metallic palladium.*

A comparison of the diffractograms in Figure 13 shows that the predominant crystalline phase of the TiO2 from the NCTiO2 support, the new catalyst and the used catalyst is tetragonal ana- tase, with diffraction angles of the respective crystal planes at: 25.1° and (101), 37° and (103), 37.5° and (112), 38° and (004), 48° and (200), 54° and (105), 55° and (211), 63.5° and (204), 67° and (116), 68° and (220), and 75° and (215), according to crystallographic chart JCPDS 21-1272. The tetragonal rutile crystalline phase of TiO2 is seen at 27°, 36°, 39°, 43° and 57°, referring to the following crystal planes (110), (101), (200), (111) and (220) respectively, according to JCPDS 21-1276 (CÁRDENAS-LIZANA et al., 2010; GAO et al., 2004; GUO et al., 2012; LI et al., 2010; LUIZ et al., 2012; WANG et al., 2011; WEHBE et al., 2009; YU et al., 2013). The diffractograms of the support and the NCTiO2-Pd5%-Sn2% catalyst are very similar, with all the peaks coinciding in position and intensity. On the other hand, this catalyst reused once and five times showed significant differences in terms of the increase in peak intensity, which indicated an increase in the crystallinity of the material. There was also the appearance of another peak at approximately 40°, which can be attributed to the greater presence of the rutile phase and/or the metallic palladium that appeared after the reduction (activation) of the catalyst *in situ* before the photocatalytic test, which shows a peak of great intensity at this diffraction angle referring to the crystalline plane (111), letter JCPDS 89-4897 (BABU et al., 2009; WANG et al., 2011; ZHANG et al., 2008).

The average crystallite size was calculated from the width at half-height of the peak corresponding to the diffraction of the anatase phase located at an angle of 27° using the Scherrer equation [14], and the values are shown in Table 9. The metal deposition practically did not alter the crystallite sizes, however, there was an increase of 13 nm from the new catalyst to the one used five times, which indicates that the repeated use and activation of the catalyst in situ can lead to a significant change in the crystalline structure towards the formation of the most stable TiO2 phase, rutile.

Table 9. Average crystallite size in nanometres of the NCTiO? support and the NCTiO2-Pd5%-Sn2% catalyst, new and reused once and five times.

Material	Average crystallite size (nm)
NCTiO2	220
NCTiO2-Pd5 %-Sn2 %	227
NCTiO2-Pd5%-Sn2% t1	228
NCTiO2-Pd5%-Sn2% t5	242

CHAPTER 3

CONCLUSION

The results presented showed that the support and catalysts supported on titanium dioxide doped with nitrogen and carbon, mainly impregnated with Pd and Sn, were very effective in the photocatalytic reduction of nitrate. Total degradation of nitrate and nitrite was achieved, with little formation of ammonium using very low concentrations of formic acid in the experiment.

The NCTiO2-Pd5%-Sn2% catalyst showed the best results due to its high photocatalytic activity, low ammonium formation, high resistance and physical-chemical stability, maintaining high nitrate conversion rates after being reused five times. It was also observed that experimental parameters such as the amount of electron donor and the spectral range of the radiation source were decisive for obtaining high photocatalytic activity, which was achieved using 15 mmol.L^{-1} of formic acid and UV-C radiation.

REFERENCES

APHA, AWWA and WEF, Standard Methods: for the examination of water and wastewater. E.W. Rice et al., Ed.; APHA, Washington, p. 1355, 2012.

BABU, N.S. et al. Influence of particle size and nature of Pd species on the hydrodechlorination of chloroaro- matics: Studies on Pd/TiO2 catalysts in chlorobenzene conversion. **Catalysis Today.** V. 141, p. 120-124, 2009.

BARBALHO, M. G. S.; CAMPOS, A. B. Natural vulnerability of soils and waters in the state of Goiás to contamination by vinasse used in the fertigation of sugarcane crops. **Boletim Goiano de Geografia.** V. 30, p. 155-170, jan./jun. 2010. Disponívelem: <http://www.revistas.ufg.br/index.php/bgg/article/view/11202/8006>. Accessed on: 08 Nov 2013.

BARBOSA, A. M.; ARAÚJO, E. S. Multi-elemental chemical analysis of water samples in municipalities in the southern mesoregion of Goiás. **Electronic Journal of the Geography Course.** Jataí-GO: Federal University of Goiás. N. 13, p. 106-122, Jul-Dec 2009. Available at: <http://revistas.jatai.ufg.br/index.php/geoambiente/article/view/968/532#.UpJKFNJHR2E>. Accessed on: 10 November 2013.

BARBOSA, D.P. et al. The use of a cation exchange resin for palladium-tin and palladium-indium catalysts for nitrate removal in water. **Journal of Molecular Catalysis A: Chemical.** V. 366, p. 294-302, 2013.

BARBOSA, D.P. **Reduction of nitrate species in water over bimetallic palladium catalysts.** 2011. 188 f. Thesis (Doctorate in Chemistry) - Institute of Chemistry, Federal University of Bahia, Salvador.

BARRABÉS, N. et al. Catalytic reduction of nitrate on Pt-Cu and Pd-Cu on active carbon using continuous reactor: the effect of copper nanoparticles. **Applied Catalysis B: Environmental.** V. 62, p. 77-85, 2006.

BARRABÉS, N.; SÁ, J. Catalytic nitrate removal from water, past, present and future perspectives. **Applied Catalysis B: Environmental.** V. 104, p. 1-5, Mar. 2011.

BOCCUZZI, F. et al. Gold, silver and copper catalysts supported on TiO2 for pure hydrogen production. **Catalysis Today.** V. 75, p. 169-175, 2002.

CAMPOS, T. S.; ROHLFS, D. B. **Evaluation of nitrate values in groundwater and their correlation with anthropic activities in the municipality of Águas Lindas de Goiás.** Postgraduate Programme in Forensic Biosciences. Institute of Pharmaceutical Studies/Pontifical Catholic University of Goiás. 2010. Available at: <http://www.cpgls.ucg.br/ArquivosUpload/1/File/V%20MOSTRA%20DE%20PRO DUO%20 CIENTIFICA/SAUDE/86.pdf>. Accessed on: 09 Nov 2013.

CÁRDENAS-LIZANA, F. et al. Gas phase hydrogenation of nitroarenes: A comparison of the catalytic action of titania supported gold and silver. **Journal of Molecular Catalysis A: Chemical.** V. 326, p. 48 - 54, 2010.

DODOUCHE, I. et al. Palladium-tin catalysts on conducting polymers for nitrate removal. **Applied Catalysis B: Environmental.** V. 93, p. 50-55, 2009.

DOLAT, D et al. One-step, hydrothermal synthesis of nitrogen, carbon co-doped titanium dioxide (N,C - TiO2) photocatalysts. Effect of alcohol degree and chain length as carbon dopant precursors on photocatalytic activity and catalyst deactivation. **Applied catalysis B: Environmental.** V. 115 - 116, p. 81 - 89, 2012.

EPRON, F. et al. Catalytic reduction of nitrate and nitrite on Pt-Cu/Al2O3 catalysts in aqueous solution: role of the interaction between copper and platinum in the reaction. **Journal of Catalysis.** V. 198, p. 309-318, 2001.

FUJISHIMA, A. et al. TiO2 photocatalysis and related surface phenomena. **Surface Science Reports.** V. 63, p. 515-582, 2008.

GAO, W. et al. Titania supported Pd-Cu bimetallic catalyst for the reduction of nitrate in drinking water. **Applied Catalysis B: Environmental.** V. 46, p. 341-351, 2003.

GAO, W. et al. Titania-supported bimetallic catalysts for photocatalytic reduction of nitrate. **Catalysis Today.** V. 90, p. 331-336, 2004.

GARRON, A. et al. Effect of the support on tin distribution in Pd-Sn/Al2O3 and Pd-Sn/SiO2 catalysts for appli- cation in water denitration. **Applied Catalysis B: Environmental.** V. 59, p. 57-69, 2005.

GUO, Y. et al. The effect of TiO2 doping on the catalytic properties of nano-Pd/SnO2 catalysts during the reduc- tion of nitrate **Applied Catalysis B: Environmetal.** V. 125, p. 21-27, 2012.

HIRAYAMA, R. et al. Combinational effect of Pt/SrTiO3:Rh photocatalyst and SnPd/Al2O3 non-photocatalyst for photocatalytic reduction of nitrate to nitrogen in water under visible light irradiation **Applied Catalysis B: Environmetal.** V. 144, p. 721-729, 2014.

HONG, J et al. Preparation of novel titania supported palladium catalysts for selective hydrogenation of acety- lene to ethylene. **Catalysis Communications.** V. 8, p. 593 - 597, 2007.

JIN, M. et al. Low temperature CO oxidation over Pd catalysts supported on highly ordered mesoporous metal oxides. **Catalysis Today.** V. 185, p. 183 - 190, 2012.

JUNG, J. et al. Nitrate reduction by maghemite supported Cu-Pd bimetallic catalyst. **Applied Catalysis B: Environmental.** V. 127, p. 148-158, 2012.

KIM, M. et al. Catalytic reduction of nitrate in water over Pd-Cu/TiO2 catalyst: effect of the strong metal-support interaction (SMSI) on the catalytic activity. **Applied Catalysis B: Environmental.** V. 142-143, p. 354361, 2013.

KOMINAMI, H. et al. Selective photocatalytic reduction of nitrate to nitrogen molecules in an aqueous suspen- sion of metal-loaded titanium(IV) oxide particles. **Chemical Communications.** p. 2933-2935, 2005.

LI, L. et al. Photocatalytic nitrate reduction over Pt-Cu/TiO2 catalysts with benzene as hole scavenger. **Journal of Photochemistry and Photobiology A: Chemistry.** V. 212, p. 113-121, 2010.

LUIZ, D.B. et al. Photocatalytic reduction of nitrate ions in water over metal-modified TiO_2. **Journal of Photo- chemistry and Photobiology A: Chemistry.** V. 246, p. 36-44, Aug. 2012.

MINISTRY OF HEALTH. Ordinance No. 2.914 of 12 December 2011. Provides for procedures to control and monitor the quality of water for human consumption and its standard of potability, Brasília, DF, Dec. 2011.

WHO. Guidelines for Drinking-Water Quality. 4ª Ed. Switzerland: World Health Organisation, 2011. p. 564. Available at:<http://whqlibdoc.who.int/publications/2011/9789241548151 eng.pdf>. Accessed on: 10 October 2013.

Opinion of the scientific committee on cosmetic products and non-food products intended for consumers: con- cerning titanium dioxide. Colipa S75, Oct. 2000. Available at: <http://ec.europa.eu/health/scientific committees/consumer safety/opinions/sccnfp opinions 97 04/index en.htm>. Accessed on: 16 Nov. 2015.

PALOMARES, A. E. et al. Nitrates removal from polluted aquifers using (Sn or Cu)/Pd catalysts in a continuous reactor. **Catalysis Today.** V. 149, p. 348-351, 2010.

PRUSSE, U.; VORLOP, K. Supported bimetallic palladium catalysts for water-phase nitrate reduction. **Journal of Molecular Catalysis A: Chemical.** V. 173, p. 313-328, 2001.

RESENDE, A. V. **Agriculture and Water Quality: Water Contamination by Nitrate.** Documents 57. Planaltina - DF: Embrapa Cerrados, 2002. p. 28. Available at: <http://ainfo.cnptia.embrapa.br/digital/bitstream/CPAC-2009/24718/1/doc 57.pdf>. Accessed: 20 Nov. 2013.

SÁ, J. et al. Can TiO2 promote the reduction of nitrates in water? **Journal of Catalysis.** V. 234, p. 282-291, 2005.

SÁ, J.; ANDERSON, J.A. FTIR study of aqueous nitrate reduction over Pd/TiO2. **Applied Catalysis B: Envi- ronmental.** V. 77, p. 409-417, 2008.

SÁ, J. et al. Photocatalytic nitrate reduction over metal modified TiO2. **Applied Catalysis B: Environmental.** V. 85, p. 192-200, 2009.

SHAND, M.; ANDERSON, J.A. Aqueous phase photocatalytic nitrate destruction using titania based materials: routes to enhanced performance and prospects for visible light activation. **Catalysis Science & Technology.** V. 3, p. 879-899, Jan. 2013.

SIKHWIVHILU, L.M. et al. Selective hydrogenation of o-chloronitrobenzene over palladium supported nano- tubular titanium dioxide derived catalysts. **Catalysis Communications.** V. 8, p. 1999-2006, 2007.

SOARES, O.S.G.P. et al. Pd-Cu and Pt-Cu catalysts supported on carbon nanotubes for nitrate reduction in water. **Industrial & Engineering Chemistry Research.** V. 49, p. 7183-7192, Jul. 2010.

aSOARES, O.S.G.P. et al. Nitrate reduction in water catalysed by Pd-Cu on different supports. **Desalination.** V. 279, p. 367-374, 2011.

SOARES, O.S.G.P. et al. Kinetic modelling of nitrate reduction catalysed by Pd-Cu supported on carbon nano- tubes. **Industrial & Engineering Chemistry Research.** V. 51, p. 4854-4860, 2012.

SOARES, O.S.G.P. et al. Photocatalytic nitrate reduction over Pd-Cu/TiO2. **Chemical Engineering Journal.** V. 251, p. 123-130, 2014.

WANG, C. et al. Rutile TiO2 nanowires on anatase TiO2 nanofibers: a branched heterostructured photocatalysis via interface-assisted fabrication approach. **Journal of Colloid And Interface Science.** V. 363, p. 157-164, 2011.

WANG, Y. et al. Remediation of actual groundwater polluted with nitrate by the catalytic reduction over copper- palladium supported on active carbon. **Applied Catalysis A: General.** V. 361, p. 123-129, 2009.

WEHBE, N. et al. Comparative study of photocatalytic and non-photocatalytic reduction of nitrates in water. **Applied Catalysis A: General.** V. 368, p. 1-8, 2009.

WEN, B. M.; LIU, C. Y.; LIU, Y. Solvothermal synthesis of ultralong single-crystalline TiO2 nanowires. **New Journal of Chemistry.** V. 29, p. 969-971, jun. 2005.

YANG, D. et al. Nitrate hydrogenation on Pd-Cu/TiO2 catalyst prepared by photo-deposition. **Catalysis Today.** V. 175, p. 356-361, 2011.

YANG, T. et al. Photocatalytic reduction of nitrate using titanium dioxide for regeneration of ion exchange brine. **Water Research.** V. 47, p. 1299-1307, 2013.

YU, X. et al. One-step ammonia hydrothermal synthesis of single crystal anatase TiO2 nanowires for highly efficient dye-sensitised solar cells. **Journal of Materials Chemistry. A.** V. 1, p. 2110-2117, 2013.

ZHANG, F. et al. High photocatalytic activity and selectivity for nitrogen in nitrate reduction on Ag/TiO2 cata- lyst with fine silver clusters. **Journal of Catalysis.** V. 232, p. 424-431, May 2005.

ZHANG, F. et al. Size-dependent hydrogenation selectivity of nitrate on Pd-Cu/TiO2 catalysts. **Journal of Phys- ical Chemical C.** V. 112, p. 7665-7671, 2008.

ZHANG, C. et al. Sodium-promoted Pd/TiO2 for catalytic oxidation of formaldehyde at ambient temperature. **Environmental Science Technology.** V. 48, p. 5816-5822, 2014.